W9-CTI-348

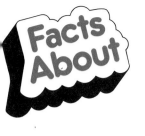

Facts About

Looking
at Stars

DONNA BAILEY

STECK-VAUGHN
LIBRARY
A Division of Steck-Vaughn Company
Austin, Texas

How to Use This Book

This book tells you many things about the stars. There is a Table of Contents on the next page. It shows you what each double page of the book is about. For example, pages 8 and 9 tell you about "Early Star Gazers."

On most of these pages you will find words that are printed in **bold** type. The bold type shows you that these words are in the Glossary on pages 46 and 47. The Glossary explains the meaning of those words that may be new to you.

At the very end of the book there is an Index. The Index tells you where to find certain words in the book. For example, you can use it to look up words like Ursa Major, Cygnes, quasars, and many other words to do with stars.

Library of Congress Cataloging-in-Publication Data

Bailey, Donna.
   Stars / written by Donna Bailey.
     p.   cm.—(Facts about)
   Summary: Examines such aspects of stars as their relationship with other heavenly bodies, their study by humanity, their chemical composition and life, and the different kinds of stars.
   ISBN 0–8114–2522–3
   1. Stars—Juvenile literature. [1. Stars.] I. Title
II. Series: Facts about (Austin, Tex.)
QB801.7.B35   1990
523.8—dc20   90-40076
                           CIP AC

Printed and bound in the United States of America
1 2 3 4 5 6 7 8 9 0 LB 95 94 93 92 91

Contents

Introduction

If you look at the sky on a very clear,
dark night you can see about 2,000 stars.
You can see thousands more if you
look at the sky through **binoculars**.
    Stars shine during the day as well,
but the bright light of the Sun hides
the light of the stars from us.

In some parts of the sky
there are so many stars
that they look like a band
of pale light.

Scientists know that each
of the stars is a ball of
glowing gas, like the Sun.
The stars look smaller
and fainter than the Sun
because they are so far away.

As you look at the stars
some of them seem to make
groups or patterns called
constellations. Some
constellations have
been named after animals.
The group in this picture
is called Ursa Major, which
means the Great Bear.

5

We live on a **planet** called Earth. It is one of nine planets that move around the Sun. The Sun and its planets make up the **Solar System.** The Solar System was made from a spinning cloud of gas and **dust.**

Our Sun is only one star in a vast group of stars called the **Galaxy.** There may be over 100 billion stars in the Galaxy. No one has been able to count all the stars. Distant stars in the Galaxy form a faint band of light across the night sky called the **Milky Way.**

Our Galaxy is only one of countless other galaxies that stretch into space. Space and everything in it is called the **Universe.**

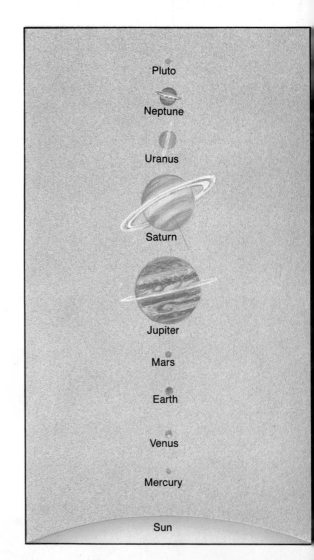

the Solar System

the arrow shows where
the Solar System is in
the Galaxy

these stars are in the
Milky Way

Early Star Gazers

Some of the first people to watch the stars were priests who believed in **astrology**. These priests noted where the stars were in the sky. They noticed that the Sun passes in front of twelve different constellations during the year. These twelve star groups are called the **zodiac** and each has its own sign.

an Egyptian coffin lid with signs of the zodiac

Arab astronomers at work 500 years ago making a map of the stars

8

William Herschel built a very large telescope to study the stars

People who study stars in a scientific way are called **astronomers**. The first astronomers wrongly thought that the Sun, planets, and stars moved around the Earth. In 1543 Nicolaus Copernicus said the Earth and other planets go around the Sun and we know now that he was right.

How We Watch the Stars

We use **telescopes** to look at stars far off in space. A telescope makes far away things look larger and closer.

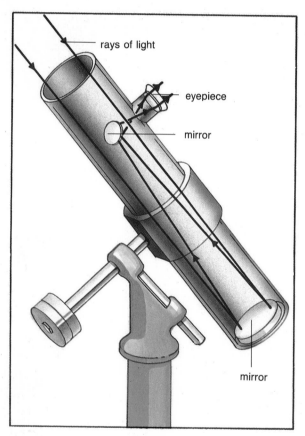

rays of light

eyepiece

mirror

mirror

a reflecting telescope uses mirrors to reflect the light and make things look larger

Rays of light enter the telescope at one end. A mirror at the back of the tube bounces the rays back to another mirror. This beams the rays to an eyepiece. In Hawaii a **reflecting telescope** has been built on top of Mauna Kea. It has a mirror 12 feet wide.

Stars send out rays of light which travel across space in a wavy motion. The distance between the top of each wave is called the **wavelength. Radio waves** are rays with the longest wavelengths. Radio waves can be picked up by big metal dishes called **radio telescopes**.

a reflecting telescope on
Mauna Kea in Hawaii

a radio telescope in
West Germany

The Northern Sky

The sky looks different depending on where you are standing on the Earth.

At the **North Pole**, a bright star called the **North Star**, or Polaris, is right over your head. The photograph, taken over a long time, shows how stars seem to move around the North Star.

The stars are not really spinning.
It is the turning of the Earth that makes
it look like the stars are moving.
Our view of the stars changes as the
Earth travels around the Sun and we see
the stars from a different **angle**.
The map shows the stars as seen from
the northern part of the world.

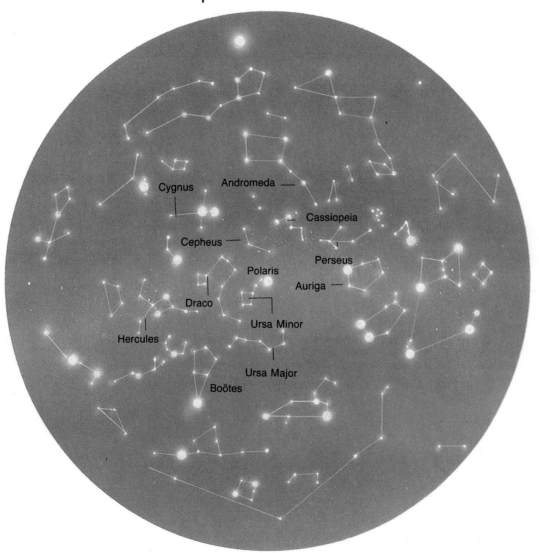

The Southern Sky

There is no bright star near the **South Pole**, but a famous cross-shaped star group called Crux helps you to find it. The photograph shows Crux against a background of thousands of stars in the southern Milky Way.

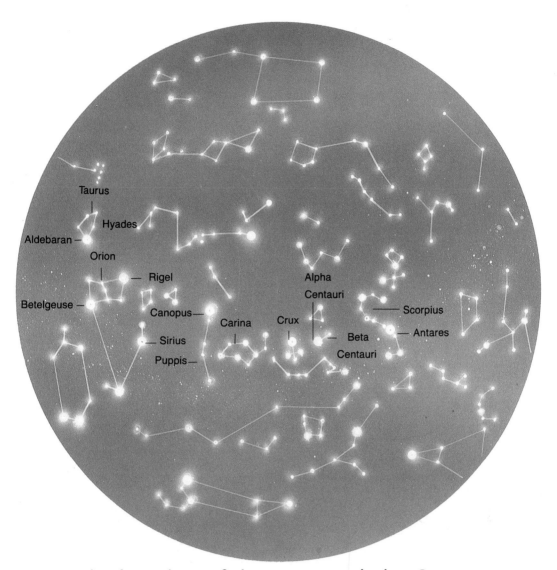

The long bar of the cross made by Crux points toward the South Pole.
Crux is the smallest constellation in the whole sky.

This map shows a view of the stars as seen from southern parts of the world. The stars at the edges can also be seen north of the **Equator**.

Star Facts

The Sun is a star, which means it is made up of gas. At the center, **hydrogen** is turned into **helium** in a **nuclear reaction**. This makes the Sun give out light, heat, and other rays. The super-hot gases rise to the surface of the Sun. Huge spouts of flame shoot out into space.

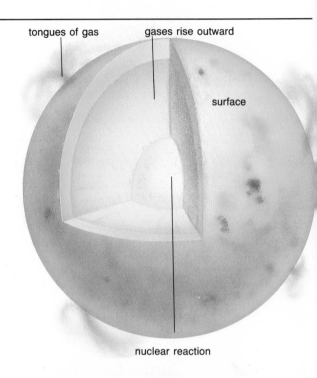

tongues of gas

gases rise outward

surface

nuclear reaction

Boötes, the Herdsman

The Sun is an average-sized star.
It is 870,000 miles across.

The constellation of Boötes, the Herdsman, has a star called Arcturus which is 27 times the size of the Sun. Arcturus is a red **giant**.

Barnard's Star is a **dwarf** in the constellation of Ophiuchus, the Snake Bearer, and is smaller than the Sun.

Barnard's star: red dwarf

Sun: yellow dwarf

Arcturus: red giant

The Life of a Star

A star is born from a huge cloud of gas and dust in space called a **nebula**. When a nebula starts to shrink and spin, **gravity** pulls the gas and dust together. The shrinking cloud is squashed at the center where it gets very hot. In the end, nuclear reactions are set off at the cloud's center. The huge ball of gas starts to shine and it gives out light and heat. A star has been born.

A star continues to turn hydrogen into helium for most of its life until it uses up all the hydrogen. Then it grows old and dies.

The photograph on page 19 shows the Great Nebula. It can be seen from Earth as a hazy blob of light in the constellation of Orion.

The photograph below shows how a star is born from a cloud of gas and dust.

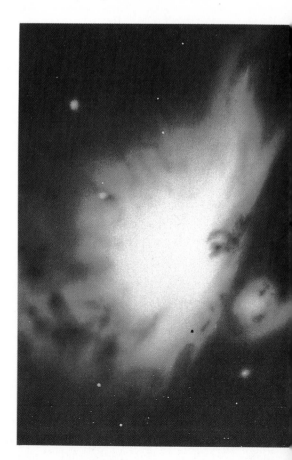

When a Star Explodes

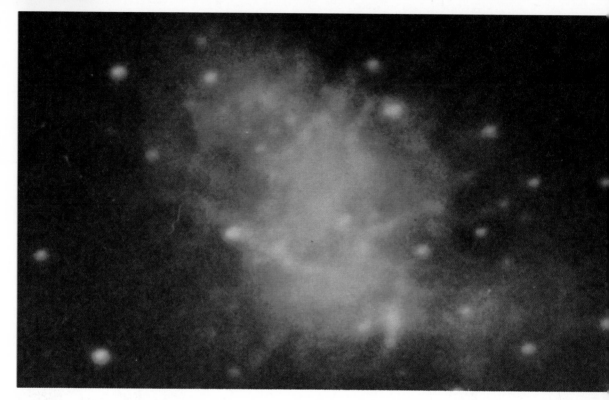

Nearly a thousand years ago a huge star blew apart and made a **supernova**. It can still be seen today as a glowing cloud of gas called the Crab Nebula.

In 1573 this drawing of a supernova was made by the astronomer Tycho Brahe.

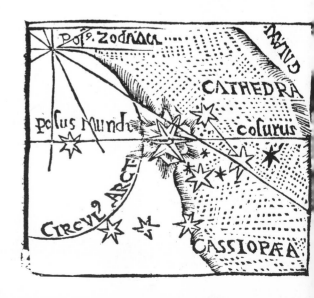

20

When a big star explodes, it produces a big flash of light. The light from a supernova can shine as brightly as one billion Suns.

The picture shows an artist's idea of how a planet would look if it was being melted down by the explosion of a huge star. Soon the planet will be completely burned up.

Clouds of Gas

The dust in a nebula is made when a star explodes.

the Horsehead Nebula shows as a dark patch in the constellation of Orion

New stars and the remains of supernovas
are strong sources of **X rays**.
We can pick up these rays by taking
photographs from **satellites** in space.
The rays show up as clouds of gas and
stars on the photographs, even when the
dust of the nebula is hiding them from
our sight.

This satellite photograph of the
Tarantula Nebula shows X rays in color
coming from a cloud of gas.

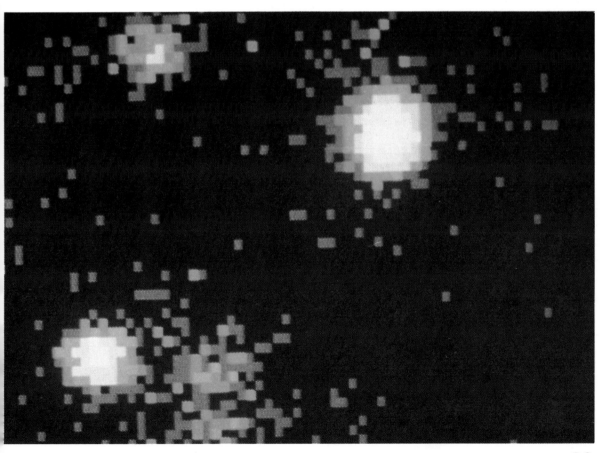

Stars with Planets

This computer photograph shows a star called Beta Pictoris which is in the constellation of Pictor. Dust and gas spin around the star. Astronomers think that after a long time this dust and gas will slowly begin to stick together in solid lumps. Over millions of years these lumps will then grow into planets.

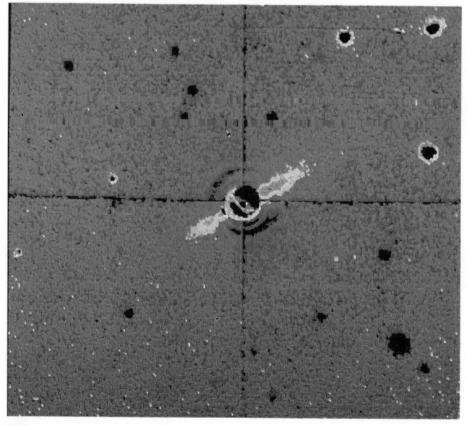

A planet like Earth takes 100 million
years to form. In our Solar System
the Earth is the best planet to live
on. This is because the Earth is at
just the right distance from the Sun
for plant and animal life.

Groups of Stars

In the constellation of Taurus you can see a group of bright hazy stars formed by a **star cluster** called the Pleiades. The stars were all formed at the same time from one cloud of gas and dust. They all travel through space together.

Most stars form in clusters at first.
The Great Nebula in Orion is producing
a cluster of hundreds of stars.

   Our own Sun may have been born in a
cluster like the Pleiades. The other stars
with the Sun have drifted away. Over
millions of years it is likely that the
stars in the Pleiades will also drift
apart.

Double Stars

When two stars are seen close together in most cases they form a **binary**.

In a binary, one star travels around the other star. Sometimes as the brighter star passes behind the fainter one its light is blocked by the star in front. The point of light we see on Earth becomes faint until the bright star comes out from behind the faint star.

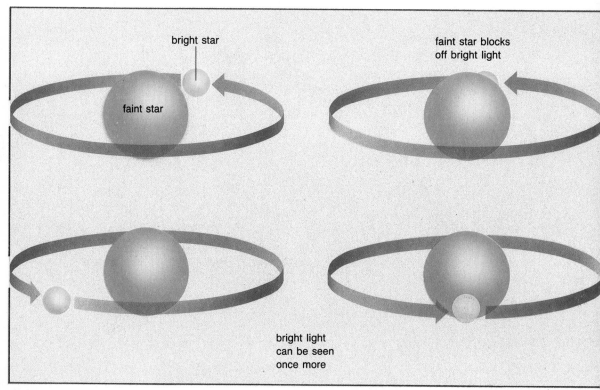

bright star

faint star blocks
off bright light

faint star

bright light
can be seen
once more

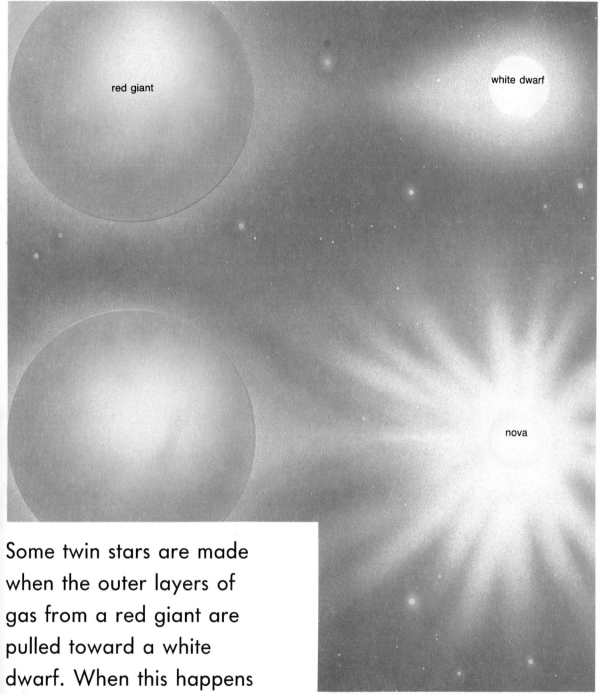

red giant

white dwarf

nova

Some twin stars are made when the outer layers of gas from a red giant are pulled toward a white dwarf. When this happens the white dwarf flares up for a few days or weeks in a blinding flash or **nova**.

A nova is different from a supernova because not all of the star blows up.

Stars that Change

Some stars can change their shapes and keep swelling up and then shrinking again. As they get bigger they get cooler and fainter. When they shrink they get brighter. These stars can take days or weeks to make these changes. At their brightest and smallest the stars can be thousands of times brighter than the Sun.

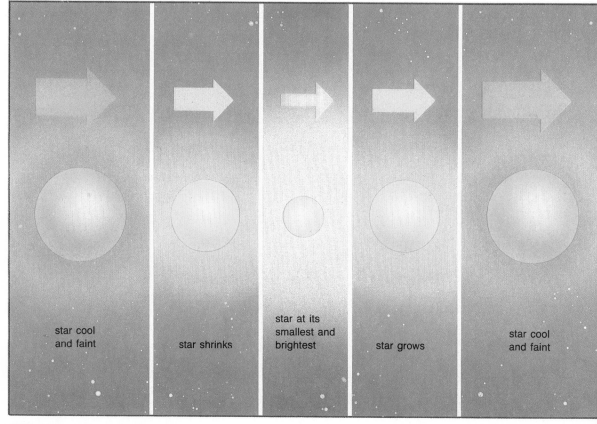

star cool and faint

star shrinks

star at its smallest and brightest

star grows

star cool and faint

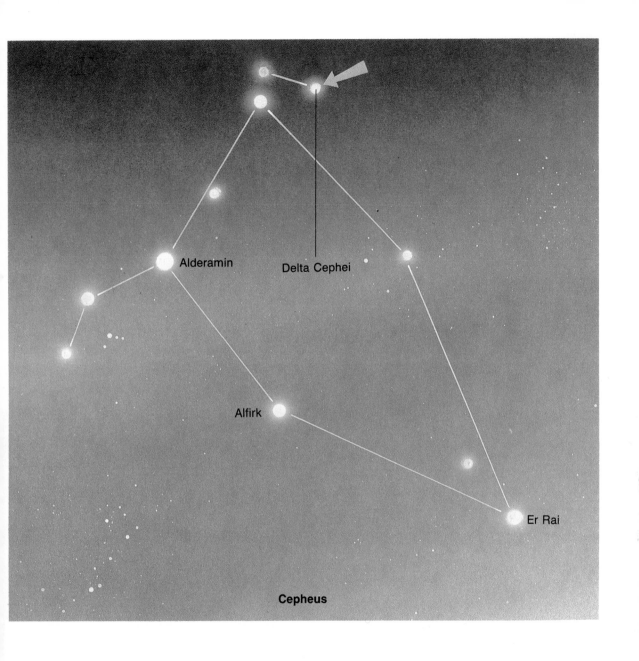

Alderamin

Delta Cephei

Alfirk

Er Rai

Cepheus

In 1976 a British astronomer, John
Goodricke, was the first person to
notice that a star called Delta Cephei
changes its shape and brightness every
five days.

Flashing Stars

When a big star blows up it sometimes leaves tiny remains called neutrons which together make a **neutron star**. There is a neutron star at the center of the Crab Nebula. It spins very quickly and gives out tiny flashes of light. These flashing stars are called **pulsars**.

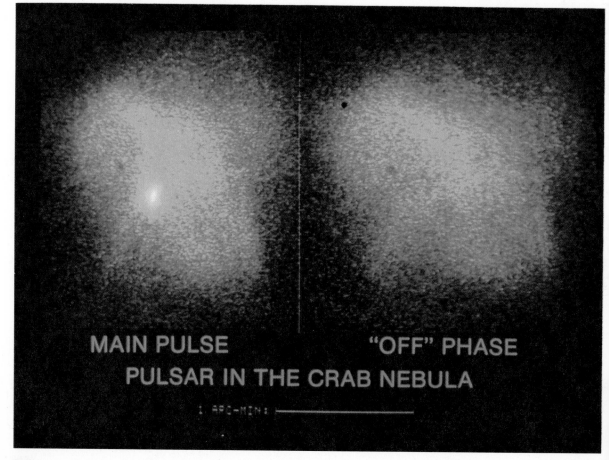

MAIN PULSE   "OFF" PHASE
PULSAR IN THE CRAB NEBULA

The youngest neutron stars are the fastest pulsars. They flash more slowly as they get older and their spin slows down. The fastest pulsars flash hundreds of times a second. The slowest ones flash once every few seconds. Not all neutron stars are pulsars.

radio telescopes on Earth can pick up the pulses of the radio waves

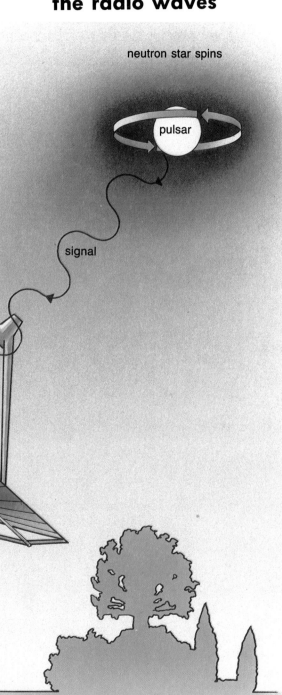

neutron star spins

pulsar

signal

radio telescope

Black Holes

Sometimes stars end their lives as a
neutron star. Sometimes very big stars
go through a further stage and
become a **black hole**. Their **core**
shrinks down until it is even smaller
than a neutron star.

The colors in this computer photograph show radio waves coming from the edge of the galaxy called Cygnus A. Scientists think the radio waves are made by gas which is being sucked into a huge black hole at the center of Cygnus A.

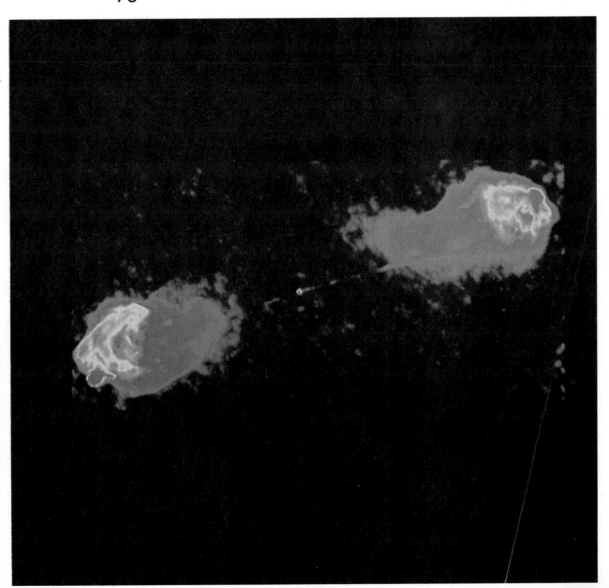

X Rays from Space

This photograph was sent back to Earth from the satellite on the right.
The photograph shows X rays coming from a **quasar** in the Virgo Cluster.

Quasars are found very far off in space in the most distant galaxies. They look like stars, but each quasar sends out as many X rays as hundreds of galaxies of stars. We have now found nearly 3,000 quasars in space but we still do not know very much about them.

How Far and How Fast?

It is difficult to tell from Earth how
far away the stars really are.
We do know that some stars are nearer
to us than other stars. Alpha Centauri
in the constellation of Centaurus the
Centaur, is nearer to us than its neighbor
Beta Centauri.

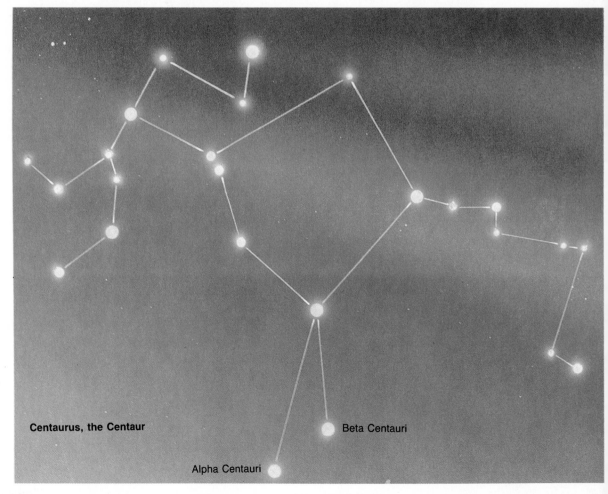

Centaurus, the Centaur

Beta Centauri

Alpha Centauri

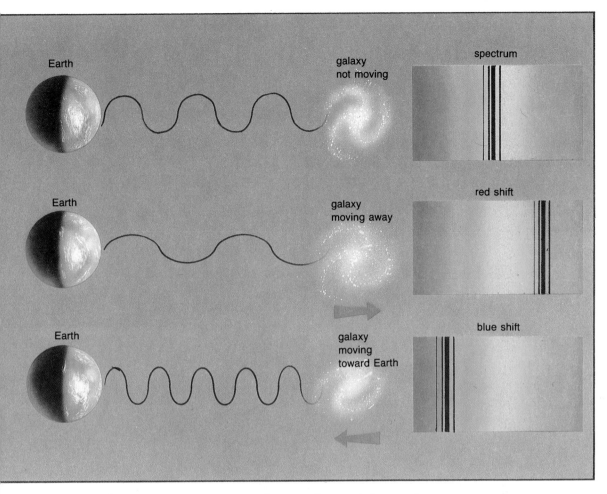

We can measure how fast things in space
are moving by using rays of light.
The light rays are broken into a
rainbow of colors called the **spectrum**.
If the star is moving away from us the
dark lines on the spectrum move toward
the red end. This is called the **red shift**.
The **blue shift** shows the object moving
toward us and the line moving toward the
blue end of the spectrum.

Galaxies

Many galaxies, such as our Milky Way, are **spiral galaxies**. Spiral galaxies have two curving arms, made up of stars and gas, which wrap around the bright center of the galaxy. The photograph shows a spiral galaxy called the Pinwheel Galaxy which is twice the size of the Milky Way.

spiral galaxy

rounded elliptical shape

irregular shape

One-fifth of all known galaxies
have a rounded shape that looks
something like an egg. These
galaxies are full of old stars.

A third kind of galaxy has no
special shape at all. We call this
an **irregular galaxy**.

How Much Do We Know?

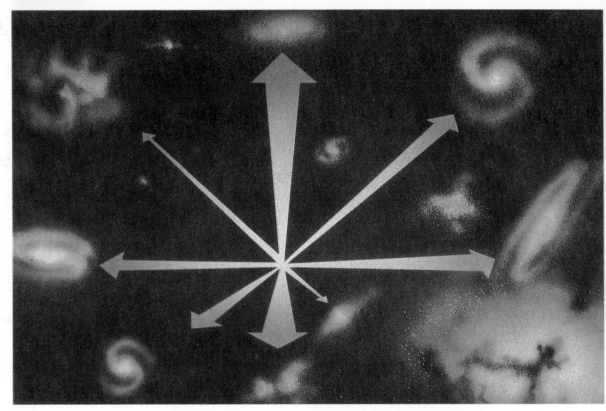

We have found out a great deal about
the Universe by looking at the stars.
We know that Earth is only a tiny
planet and that our Sun is not at the
center of the Universe.

We are only a tiny part of our Galaxy
and there are countless other galaxies
like it far away in space. Most
galaxies are moving away from us.

The picture below shows a huge mirror being made ready for use in the Hubble Space Telescope.

The telescope will be operating outside the Earth's atmosphere making objects appear 100 times clearer than ever before which is something not even the best telescopes on Earth can do. It was sent into space in 1990.

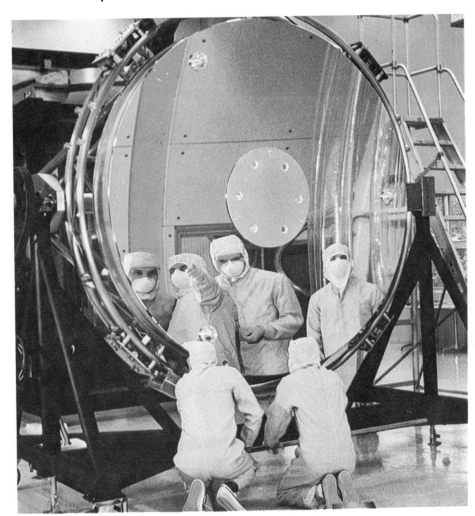

the Hubble Space Telescope

Travel to the Stars

People on Earth often dream of traveling to the stars. There are many stories and movies about space travel and exploring space. In the future we may build space ships like the one in this picture.

Travel to the stars will not be easy. First we must send **space probes** to the stars, like the ones sent to planets in the Solar System. However, even our nearest star is 300,000 times farther away than the Sun. This means that we will need spacecraft that can go much faster than today's spacecraft.

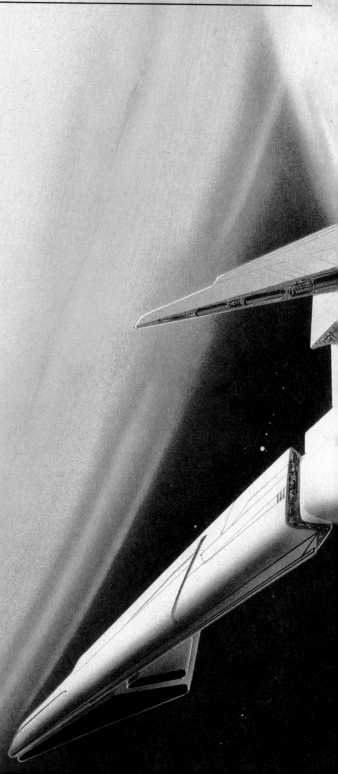

Glossary

angle the point at which one line joins another line.

astrology the belief that the position of stars and planets have the ability to affect people's lives.

astronomer someone who studies the stars and planets.

binary a double star system in which one star goes around the other.

binoculars a kind of double telescope with two eyepieces.

black hole the remains of a star. It has a very strong pull of gravity and sucks in every object around it.

blue shift a movement of lines in the spectrum of light toward the blue end. It shows that an object in space is moving toward us.

constellation a group of stars as they are seen from Earth.

core the very center of something.

dust tiny pieces of solid matter.

dwarf a very small star, about 100 times smaller than the Sun.

Equator the imaginary line around the center of the Earth.

galaxy a group of millions and millions of stars, all loosely held together by gravity. Our own Galaxy is called the Milky Way.

giant a very large star, much bigger than the Sun.

gravity the force that pulls objects toward each other.

helium a gas found in stars and out in space.

hydrogen a very light gas found in stars and out in space.

irregular galaxy a galaxy that has no special shape.

Milky Way a band of hazy gas and stars we can see in the night sky.

nebula a cloud of gas and dust in space.

neutron star the squashed core of a star that has blown up. It is very small and spins very fast.

North Pole the farthest point north on a planet.

North Star brightest, readily visible star located above the North Pole. It is also called the polestar.

nova a huge flash of light from a double star system that contains a white dwarf.

nuclear reaction a change that takes place inside a star and makes the star shine.

planet a large object that moves around a star. The Earth is a planet that moves around the Sun.

pulsar the remains of a dead star that gives out a regular pulse of radio waves as it spins.

quasar an object far away in space that looks like a star but has much more energy.

radio telescope a telescope made to pick up radio waves instead of light.

radio wave a ray that has the longest wavelength of all.

red shift a movement of the lines in the spectrum toward the red end. It shows that a star is moving away from us.

reflecting telescope a kind of telescope that works with mirrors.

satellite an object in space that moves around another object. Spacecraft that go around the Earth are satellites.

Solar System the Sun and all the objects that go around it.

South Pole the farthest point south on a planet.

space probe a machine with no people inside that is sent from Earth to study objects in space.

spectrum a range of light-rays or any other kind of rays.

spiral galaxy a kind of galaxy that has a flattened shape and arms curving out from the center.

star cluster a group of stars that are very close together.

supernova the explosion of a very big star at the end of its life.

telescope an instrument that is used for looking at distant objects or for picking up the rays they give out.

Universe all of space and everything in it.

wavelength the distance between the top of one wave and the top of the next. It is how we measuare different kinds of rays, such as light, which travel in a wavy motion.

X ray a type of ray that has a very short wavelength. X rays are given out by very hot gases in space.

zodiac the twelve constellations in the sky.

Index

© Heinemann Children's Reference 1990
Artwork © BLA Publishing Limited 1987
Material used in this book first appeared in
Macmillan World Library; *The Stars.*
Published by Heinemann Children's
Reference.